Allen Francis Wood

Wood's Outline Astronomy

The Last of a Graded Series of Outlines Including Botany, Physiology,

Physics, Meteorology and Astronomy

Allen Francis Wood

Wood's Outline Astronomy
The Last of a Graded Series of Outlines Including Botany, Physiology, Physics, Meteorology and Astronomy

ISBN/EAN: 9783337779184

Printed in Europe, USA, Canada, Australia, Japan

Cover: Foto ©berggeist007 / pixelio.de

More available books at **www.hansebooks.com**

WOOD'S
OUTLINE ASTRONOMY:

THE LAST OF A GRADED SERIES OF OUTLINES,

INCLUDING

BOTANY, PHYSIOLOGY, PHYSICS, METEOROLOGY, AND ASTRONOMY.

PREPARED EXPRESSLY FOR USE IN HIS OWN SCHOOL,

BY

ALLEN F. WOOD, A. M.,

MASTER OF FIFTH STREET GRAMMAR SCHOOL,

NEW BEDFORD, MASS.

SUITED TO USE IN AVERAGE GRAMMAR SCHOOLS, AND
IN HIGH SCHOOLS WITH CLASSICAL SCHOLARS
WHOSE TIME FOR THE SCIENCES
IS LIMITED.

———

NEW BEDFORD:
E. ANTHONY & SONS, PRINTERS.
1877.

PREFACE.

IN the progress of education, the outlines of some of the sciences seem to demand a place in our Grammar schools. The elements of Botany, Physiology, Physics, Meteorology, and Astronomy, may be made as useful as Geography and History, and unless they are taught in our Grammar schools, more than eighty per cent. of our children have no opportunity of gaining any accurate knowledge of them.

The *objections* which are commonly urged against these elements, or outlines, are mainly two: first, that there is no time in the Grammar schools to give to them, and second, that the scholars are not able to comprehend them.

In regard to the *first* of these objections, I remark that only two hours a week for the major portion of a year, is all that is necessary to complete the outlines of any one of the sciences named, and a four years experience shows that this amount of time can be devoted to these subjects, without seriously affecting the ordinary studies.

In regard to the *second* objection, that the scholars are not capable of comprehending these sciences, I reply that when the facts are properly presented, as large a proportion of the scholars comprehend them, as comprehend Arithmetic, Geography, or any other study.

There is another point which is sometimes urged as an objection to these sciences in the Grammar schools, viz., that the High school is a more fitting place for them; but when we consider that only fifteen per cent. of the scholars that enter the Grammar schools ever enter the High school, the argument is largely in favor of giving some attention to these sciences in the Grammar schools.

The *order* in which these sciences should be arranged for our schools is that in which they were enumerated, viz., Botany, Physiology, Physics, Meteorology, and Astronomy. The principal points of Botany can be so stated that average children ten years old can understand them; and the outlines of the other sciences can be introduced, one each year, as the children advance in their course of study.

In order to make the study of these sciences a *success* in the Grammar schools, two facts must be kept constantly in mind : *first*, the most important topics in each science must be chosen, and *second*, these topics must be presented in the clearest possible manner.

To meet both of these demands *without a suitable manual*, is no easy task, and, in addition to the work ordinarily required, imposes a heavy burden upon the teacher; for she is obliged, as she advances with her class, not only to prepare a new outline each year, but to see that it is accurately copied by all her scholars.

With the hope of relieving the teachers immediately associated with himself, and also of securing more uniform methods of teaching the sciences in question, the writer has undertaken to prepare a brief outline of each. These outlines will be adapted to the different grades of his school, and it is expected that they will all be ready for use before the close of the present school year.

If these outlines should seem to meet the wants of other teachers, and should be introduced into other schools than that for which they were prepared, every effort to lift the burden of teachers, and at the same time to diffuse a knowledge of the sciences among the children, will be doubly repaid.

The Outline Astronomy is now ready for use. It is designed for scholars only fourteen years of age, and therefore is not a complete treatise. If any of the topics need further development, additions may be made upon the blank pages which have been inserted for the convenience of the teacher.

The authorities which have been consulted in the preparation of this outline are Lockyer's Elements of Astronomy, Quackenbos's Natural Philosophy, and Appleton's New Cyclopedia.

In using the book, every teacher must exercise his own judgment as to the best methods of teaching the subject. In his own practice, the author treats every topic *orally*, when it is brought before the class for the first time. In doing so, he carefully explains all the terms and principles, and illustrates them by all the appliances at his command. Then, when all the terms and principles are well understood, he requires his scholars to fix the outlines firmly in mind, and reviews them often.

A. F. W.

NEW BEDFORD, Dec. 19, 1876.

ASTRONOMY.

INTRODUCTION.

1. **Definition.**—Astronomy is the science that treats of the heavenly bodies, viz., the sun, moon, stars, planets, and comets.

2. **Number of Heavenly Bodies.**—The number of the heavenly bodies is not known, and never can be.

3. **Shape.**—The shape of these bodies, or *worlds* as they are often called, is generally round.

4. **Visibility.**—Very many of these bodies shed a light of their own, or a borrowed light, and are therefore visible to the human eye. Others are non-luminous, and consequently invisible.

5. **Groups or Systems.**—All the bodies that occupy space are arranged in groups or systems, and the members of each system take different names according to their peculiarities.

6. **The Sun of each System.**—The most important member of each system is its *sun*. It occupies a central position in the system, and is the source of light and heat to the other members. It contains more matter than all the other members of the system, and by its attractive force holds them in their places.

1

7. **Planets Defined.**—The planets in each system are those bodies that revolve about the *sun* of the system, and shine by its reflected light.

8. **Moons.**—Moons are large round bodies that move round the planets in elliptical orbits, and shine by the reflected light of the sun.

9. **Stars.**—Stars are bright luminous bodies seen in every part of the heavens at night.

10. **Comets.**—Comets are luminous heavenly bodies which usually present long tails of light.

SOLAR SYSTEM.

11. **The Earth a Member.**—The Earth which we inhabit is a planet belonging to the Solar System, of which the Sun is the centre.

12. **Its Members.**—The Solar System, as known at present, consists of the Sun, 138 planets, 18 moons, and thousands of comets.

THE SUN.

13. **Size.**—The Sun is a round body, 853,000 miles in diameter. It contains many times as much matter as all the other members of the solar system taken together, and by its attractive force holds them in their places.

14. **Constitution.**—Nothing is known about the nature of the matter that makes up the great interior bulk of the Sun. In its atmosphere, the following substances have been found to exist in the form of vapor:

sodium, calcium, magnesium, iron, zinc, copper, and nickel.

15. **Motions.**—The Sun has two motions, one on its axis in 25 days, and the other in an elliptical orbit in 18,000,000 years. In this latter movement the Sun, with all the other members of the solar system, moves through space at the rate of 300 miles a minute.

16. **Light.**—The surface of the Sun is almost wholly luminous, and diffuses a bright light to all the planets. The intensity of the light at the different planets varies according to their distances from the Sun.

17. **Spots.**—Dark spots are at times seen upon the surface of the Sun. They sometimes continue for months, sometimes only for a day or two, and they have been known to appear and then to disappear almost instantly.

18. **Heat.**—The Sun is the great source of heat to all the planets. The intensity of the heat at the different planets varies according to their distances from the Sun.

PLANETS.

19. **Definition.**—The planets of the solar system are those bodies that move around our Sun in elliptical orbits, and shine by its reflected light.

20. **Number.**—In the solar system there are known to be 138 planets. Of these, 130 are so small that they are called asteroids, or star-like bodies.

21. **Size.**—The asteroids vary in their diameters

from 17 to 228 miles; the other planets, from 3,000 to 37,000 miles.

22. A Planet's Distance from the Sun Variable.—The orbit being elliptical, a planet's distance from the Sun is constantly varying. When a planet's distance from the Sun is spoken of, its mean or average distance is meant. This is obtained by adding its greatest distance from the Sun to its least distance, and dividing by 2.

23. Perihelion and Aphelion.—That point of a planet's orbit which lies nearest to the Sun is called its perihelion; and the point farthest distant is called its aphelion.

24. Distances from the Sun.—The planets are not all equally distant from the Sun, but vary from 35 millions to 2750 millions of miles.

25. Motions.—Every planet has two motions, one on its axis, and the other round the Sun.

26. Day and Year.—The time it takes a planet to turn once on its axis is called its *day*, and the time it takes for it to go once round the Sun is called its *year*.

27. Light.—The planets shine by a steady reflected light of the Sun.

28. Transit.—The passage of a smaller body across the disk of a larger is called its transit. Thus the passages of Mercury and Venus across the Sun's disk, are called, respectively, the Transit of Mercury and the Transit of Venus.

29. Inferior and Superior.—Mercury and Venus are nearer the Sun than the Earth is, and are called

Inferior Planets. The others are farther from the Sun than the Earth is, and are called Superior Planets.

30. **Names.**—The following are the names of the planets in the order of their distances from the Sun: Mercury, Venus, Earth, Mars, Asteroids, Jupiter, Saturn, Uranus, and Neptune.

The planets will now be considered in the above order.

MERCURY.

31. **Position.**—Mercury is the nearest planet to the Sun.

32. **Visibility.**—It can be seen only a few times each year. When visible, it is just after sunset, or just before sunrise. At other times it is lost below the horizon, or in the brightness of the Sun.

33. **Light.**—Unlike the other planets, it has a slight twinkle to its light.

34. **Solar Light and Heat.**—The Sun's light and heat at Mercury are seven times as intense as they are at the Earth. This fact alone makes it impossible for Mercury to be inhabited by creatures like ourselves.

35. **Day and Year.**—Mercury's day varies but a few minutes from our own, while its year is equal to less than three of our months.

VENUS.

36. **Position.**—Venus is the second planet from the Sun, and in its course comes nearer the Earth than any other of the planets.

37. **Visibility.**—Venus is so bright that it is sometimes seen with the naked eye at midday. During parts of the year it rises before the Sun, and then is called the Morning Star; at other times it rises after the Sun, and then is called the Evening Star.

38. **Solar Light and Heat.**—The Sun's light and heat at Venus are twice as great as they are at the Earth, and they vary much more between the equator and the poles.

39. **Day and Year.**—Venus's day is about half an hour less than our own, and its year is a little less than eight of our months.

THE EARTH.

40. **Position.**—The Earth on which we live is the third planet from the Sun.

41. **Form.**—The Earth is nearly a perfect sphere, being flattened a little on opposite sides. The highest mountains are so slight when compared with the great mass of the Earth. that they do not materially affect its shape.

42. **Diameter.**—The mean diameter of the Earth is 7912 miles.

43. **Surface.**—The surface of the Earth is composed of land and water, one fourth of it being land and three fourths being water.

44. **Constitution.**—According to present discoveries, the Earth is made up of 64 substances or elements. The most common of these are the metals,—gold, silver, iron, copper, mercury, lead, tin, antimony, zinc,

bismuth, arsenic, and nickel; and the gases,—oxygen, hydrogen, and nitrogen.

45. Two Motions.—The Earth has two motions, one on its axis in 24 hours, and the other round the Sun in 365¼ days. The former is called its *daily* motion, and the latter its *yearly*.

46. Day and Night.—From the Earth's daily motion, we have the succession of day and night. It occurs in this way. During the 24 hours, every part of the Earth, except small portions at the poles, comes into the Sun's light, and then recedes from it. While the Sun shines upon any place, it is *day* there; during the rest of the 24 hours, the place is withdrawn from the Sun's rays, darkness reigns, and it is *night* there.

47. Orbit.—The orbit of the Earth is really an ellipse, like that of the other planets, but it deviates only a little from a circle.

48. Distance from the Sun.—The Earth's perihelion is 90,000,000 miles from the Sun, and its aphelion 93,000,000 miles. During the year, the Earth's distance from the Sun varies between these two distances. The mean distance may be stated at about 92,000,000 miles.

49. Change of Seasons.—The position of the Earth is such while moving round in its orbit, that the Sun's vertical rays strike the Earth north of the equator during half the year, and south of the equator during the rest of the year. When the Sun reaches its most northern point in the heavens, June 21, it gives the most heat to the northern hemisphere, and we have summer; at the same time it gives the least heat to the southern hemisphere, and it is winter there. Six months later, December 21, the Sun is at its most southern point, and gives the most heat to the southern hemisphere. It is then summer south of the equator, while it is winter here. In either hemisphere spring follows winter, and autumn, or fall, follows summer.

50. **Position Designated.**—The position of a place on the Earth is designated by its latitude and longitude. Thus New Bedford is in latitude 41 degrees 38 minutes North, and longitude 70 degrees 55 minutes West.

51. **Significance of the Tropics.**—These two circles 23½ deg. north and south from the equator, mark the space beyond which the Sun's rays are never vertical. They are called tropics, because when the Sun reaches them, in its apparant movement about the Earth, it turns back towards the equator. The northern tropic is called the Tropic of Cancer, because, at the point of turning, the Sun is seen in that portion of the heavens which is occupied by a cluster of stars called Cancer. For a like reason, the southern tropic is called the Tropic of Capricorn.

52. **Polar Circles.**—The Arctic circle marks the coldest portion of the northern hemisphere. It surrounds the north pole at a distance of 23½ deg., and takes its name from the fact that it lies under the constellation formerly called Arcticus, but now called by us the Bear. The Antarctic circle surrounds the south pole, as the Arctic does the north, at a distance of 23½ deg. It marks the coldest portion of the southern

hemisphere, and takes its name from the fact that it lies opposite to the Arctic circle.

53. Climate, etc.—Climate and vegetation vary greatly between the equator and the poles, owing largely to the difference in direction of the Sun's rays.

54. Direction of the Earth's Axis.—The Earth's axis always points to a place in the heavens near the North Star.

55. Horizons.—The Sensible Horizon is the circle on the Earth's surface which bounds our view. The Rational Horizon is a plane passing through the centre of the Earth parallel to the sensible horizon, and extending out into the heavens.

56. Zenith—Nadir.—The Zenith is the point in the heavens directly over our heads. The Nadir is the point in the heavens directly under our feet, or in other words, it is the point in the heavens directly over the heads of our antipodes.

57. Equinoctial.—The Equinoctial, or Celestial Equator, is the plane of the Earth's equator, extending to the heavens and dividing *them* as it does the Earth.

58. Ecliptic.—The Ecliptic is the apparent path of the Sun. It is a great circle of the heavens, and cuts the equinoctial, or celestial equator, at an angle of 23½ degrees.

59. Equinoxes.—The two points at which the ecliptic cuts the equinoctial are called Equinoxes, because when the Sun reaches these points in March and September, the days and nights are equal all over the world.

B

60. **Solstices.**—The Solstices are the standing-points of the Sun. When the Sun reaches its most northern limit, June 21st, it appears for several days to stand still, that is to move neither north nor south, and is then said to be at its Summer Solstice. On December 21st, the Sun reaches its other standing-point, and is then said to be at its Winter Solstice.

61. **Zodiac.**—The Zodiac is an imaginary belt in the heavens 18 deg. wide, 9 deg. of which lie on each side of the ecliptic. It is in the Zodiac that we see the Sun, Moon, and all the greater planets. The entire belt of the Zodiac is divided into 12 signs of 30 deg. each.

62. **Signs of the Zodiac.**—The signs of the Zodiac are clusters of stars named as follows:

Aries, the ram.	Libra, the balance.
Taurus, the bull.	Scorpio, the scorpion.
Gemini, the twins.	Sagittarius, the archer.
Cancer, the crab.	Capricornus, the goat.
Leo, the lion.	Aquarius, the water-bearer.
Virgo, the virgin.	Pisces, the fishes.

63. **Moon.**—The Earth is attended by one moon.

64. **Size—Distance.**—The Moon is 2160 miles in diameter, and distant from the Earth 239,000 miles.

65. **Orbit.**—The Moon's orbit is elliptical.

66. **Perigee—Apogee.**—When the Moon is nearest to the Earth it is said to be in perigee, and when farthest from it, in apogee.

67. **Motions.**—The Moon turns on its axis in exactly the same time that it takes it to go once round the Earth. These motions are completed in about 27 days.

68. **Light.**—The Moon shines only by the reflected light of the Sun, and is visible at night only about half of each month.

69. **Phases of the Moon.**—While the Moon is revolving about the Earth, the bright surface presented to us is constantly changing. Hence arise the Phases of the Moon.

MARS.

70. **Position.**—Mars is the fourth planet in order of distance from the Sun, and is nearer the Earth than any other of the superior planets.

71. **Appearance.**—Under the *telescope*, it sheds a white light from its poles, and red and green tints from other portions of its surface; but to the *naked eye*, it presents a *reddish* light. The white light is supposed to be reflected from its snow and ice, the red light from its soil, and the green light from its oceans.

72. **Size.**—Mars is only one sixth as large as the Earth.

73. **Light and Heat.**—As Mars is farther from the Sun than the Earth is, its climate is colder than that of the Earth, and its light is less intense.

74. **Day and Year.**—Its day is about the same as ours, and its year is nearly twice as long.

ASTEROIDS.

75. Position.—The Asteroids are scattered in space between Mars and Jupiter.

76. Number.—It is known that there are 130 Asteroids, and it is believed that there are very many more.

77. Visibility.—With the exception of Ceres and Vesta, the Asteroids are invisible to the naked eye.

78. Discovery.—All the Asteroids which are now known have been discovered since 1801.

79. Size.—The Asteroids are the smallest of the known planets, Vesta, the largest, having a diameter of only 228 miles.

80. Attraction.—The power of attraction at the Asteroids is very slight. A large man would weigh less than one pound, and a large horse would weigh less than ten pounds.

JUPITER.

81. Position.—Jupiter lies beyond the Asteroids, and with the exception of these minute planets, it is the fifth in order of distance from the Sun.

82. Size.—Jupiter is the largest planet, and is 1200 times as large as the Earth.

83. Day and Year.—Jupiter's day is about 10 hours long, and its year is equal to 12 of ours.

84. Moons.—Jupiter is attended by four satellites, or moons, three of which are larger than our own. They are invisible to the naked eye, and it is seldom

that their position is such that they can all be seen with a telescope at the same time.

SATURN.

85. **Position.**—Saturn lies next to Jupiter, and of the larger planets is sixth in order of distance from the Sun.

86. **Size.**—Saturn is next to Jupiter in size, and is 700 times as large as the Earth.

87. **Appearance.**—Saturn is not itself remarkably brilliant, but is surrounded by three bright rings, made up, it is believed, of minute satellites.

88. **Day and Year.**—Saturn's day is about 10½ hours long, and its year is equal to 29½ of ours.

89. **Moons.**—Saturn is attended by eight moons, one of which, Titan, is the largest in the solar system.

URANUS.

90. **Position.**—With the exception of the Asteroids, Uranus is the seventh planet in order of distance from the Sun, and is the outermost, but one, of all the known members of the planetary system.

91. **Size.**—Its volume is 74 times that of the Earth.

92. **Appearance.**—Owing to the great elongation of its orbit, its brightness varies considerably during its year.

93. **Day and Year.**—Its day is very uncertain, but is thought to be about 9½ hours long. Its year is equal to 84 of ours.

94. **Moons.**—Uranus is attended by four moons.

B2

NEPTUNE.

95. **Position.**—Neptune is the most distant of all the known planets, and, with the exception of the Asteroids, it is the eighth in order of distance from the Sun.

96. **Size.**—Its volume is 105 times that of the Earth.

97. **Visibility.**—Neptune is nearly three billion miles from the Earth, is invisible to the naked eye, and was not discovered by the aid of the most powerful telescopes till 1846.

98. **Day and Year.**—The length of Neptune's day has not been determined, and perhaps never will be, inasmuch as no spots have been discovered on its surface by which the period of its revolution can be ascertained. Its year is equal to 165 of ours.

99. **Moons.**—Neptune is attended by at least one moon.

ECLIPSES.

100. **Definition.**—An eclipse of the Sun or Moon is its temporary obscuration by the intervention of some other body. When the whole disk is darkened, the eclipse is said to be total; and when only a portion of the disk is darkened, it is said to be partial.

101. **Eclipse of the Sun.**—An eclipse of the Sun is caused by the passage of the Moon between the Sun and the Earth, thus cutting off a part of the Sun's direct rays to the Earth.

102. **Eclipse of the Moon.**—An eclipse of the Moon is caused by the passage of the Earth between

the Sun and the Moon, thus cutting off a part of the Sun's direct rays to the Moon, and leaving a part of the Moon in darkness.

103. **Number of Eclipses.**—The Sun and Moon may together have seven eclipses in a year, and cannot have less than two. They usually have four.

COMETS.

104. **Position.**—The comets are scattered at a great variety of distances from the Sun, some coming within the orbits of the planets, while others lie far beyond.

105. **Orbits.**—The orbits of the comets are either ellipses, parabolas, or hyperbolas. If a comet has an elliptical orbit, it revolves around the Sun in regular intervals; if it describes a parabola, or hyperbola, it will in time leave the solar system and seek other companions millions of miles away.

106. **Appearance.**—A comet, in its most common form, presents three quite distinct parts. The brightest of all is quite small when compared with the others, and is called the *nucleus*. Surrounding the nucleus is a dimmer part, called the *coma*, or *head;* and streaming from this, like hair in the wind, is a *tail of light*, which assumes a great variety of shapes.

107. **Light.**—The comets shine by their own light. This fact is accounted for on the ground that the comets, unlike the planets, are all white-hot.

108. **Number.**—It is thought that there are millions of comets belonging to the solar system, though less than a thousand have been actually recorded.

STARS.

109. Position.—The stars are scattered everywhere in space outside the solar system, and, unlike the planets, they always seem to occupy the same positions with reference to each other.

110. Magnitudes.—All the stars that are visible to the naked eye, are divided, according to their brightness, into six classes, called magnitudes. The brightest stars are said to be of the *first* magnitude, the next brightest of the *second* magnitude, and so on through the *third*, *fourth*, *fifth*, and *sixth* magnitudes, the *last* being so faint as scarcely to be visible. If we introduce the telescope, other stars, called Telescopic stars, are brought to view. These are arranged as *seventh*, *eighth*, and other magnitudes, to the twentieth.

111. Number.—The number of stars visible to the naked eye does not much exceed 6000, and the number visible at any one time does not exceed 3000. The number of stars that can be seen through a telescope is not far from 20,000,000.

112. Distances.—The stars are all situated at immense distances from us. The nearest star is 20 trillions of miles away, and it took its light, travelling at the rate of 185,000 miles a second, three and a half years to reach the Earth. The light of some stars required more than 4000 years to reach the Earth, and if these stars were extinguished to-day, we should enjoy their light for the same period of time.

113. Light.—The stars all shine by their own light,

and it is probable that they are suns to systems of their own.

114. **Movement.**—The stars, in common with our Sun, move around some enormous body trillions of miles away in space.

115. **Constellations.**—The stars are divided into 109 constellations, which take their names from animals and other objects to which their outline bears some imaginary resemblance.

116. **Galaxy.**—The Galaxy, or Milky Way, is an irregular band of light passing nearly over our heads in the heavens. It is composed of innumerable stars, which are so distant as to be distinguished only by the aid of the most powerful telescopes.

TABLE OF CONTENTS.